T0332447

Scientific Methods Used in Research and Writing

Advances in Mathematics and Engineering

Series Editor: Mangey Ram, Department of Mathematics, Graphic Era University, Dehradun, Uttarakhand, India

The main aim of this focus book series is to publish original articles that bring up the latest development and research in mathematics and its applications. The books in this series are in short form, ranging between 20,000 and 50,000 words or 100 to 125 printed pages, and encompass a comprehensive range of mathematics and engineering areas. It will include, but won't be limited to, mathematical engineering sciences, engineering and technology, physical sciences, numerical and computational sciences, space sciences, and meteorology. The books within this series will provide professionals, researchers, educators, and advanced students in the field with an invaluable reference into the latest research and developments.

Recent Advancements in Software Reliability Assurance
Edited by Adarsh Anand and Mangey Ram

Linear Transformation
Examples and Solutions
Nita H. Shah and Urmila B. Chaudhari

Scientific Methods Used in Research and Writing
Mangey Ram, Om Prakash Nautiyal, and Durgesh Pant

Journey from Natural Numbers to Complex Numbers
Nita Shah and Thakkar D. Vishnuprasad

For more information about this series, please visit: https://www.crcpress.com/Advances-in-Mathematics-and-Engineering/book-series/AME

Scientific Methods Used in Research and Writing

Edited By

Mangey Ram
Om Prakash Nautiyal
Durgesh Pant

CRC Press is an imprint of the
Taylor & Francis Group, an **informa** business

First edition published 2021
by CRC Press
6000 Broken Sound Parkway NW, Suite 300, Boca Raton, FL 33487-2742

and by CRC Press
2 Park Square, Milton Park, Abingdon, Oxon, OX14 4RN

Library of Congress Cataloging-in-Publication Data

Names: Ram, Mangey, editor. | Nautiyal, Om Prakash, editor. | Pant, Durgesh, editor.
Title: Scientific methods used in research and writing / edited by Mangey Ram, Om Prakash Nautiyal, and Durgesh Pant.
Description: Boca Raton : CRC Press, 2021. | Series: Advances in mathematics and engineering | Includes bibliographical references and index.
Identifiers: LCCN 2020034869 (print) | LCCN 2020034870 (ebook) | ISBN 9780367627140 (hardback) | ISBN 9781003119180 (ebook)
Subjects: LCSH: Research. | Academic writing.
Classification: LCC Q179.9 .S358 2021 (print) | LCC Q179.9 (ebook) | DDC 001.4/22--dc23
LC record available at https://lccn.loc.gov/2020034869
LC ebook record available at https://lccn.loc.gov/2020034870

ISBN: 978-0-367-62714-0 (hbk)
ISBN: 978-1-003-11918-0 (ebk)

Typeset in Times
by SPi Global, India

Contents

Preface

Nowadays research publications, research projects, and teaching learning have become the most important part of universities, institutions, organizations, and industries. Day-to-day, a lot of research articles, book chapters, monographs, and research projects are submitted to the journals, books, and funding agencies, but most of them are rejected due to lack of basic academic writing skills, scientific methods. This book, *Scientific Methods Used in Research and Writing*, gives the basic staring idea to the researcher or teacher about the research article writing skills, statistical methods in scientific research, teaching-learning theories and the development of research projects.

Mangey Ram
Graphic Era (Deemed to be University), India

Om Prakash Nautiyal
Uttarakhand Science Education & Research Centre, India

Durgesh Pant
Uttarakhand Science Education & Research Centre, India

Preface

Acknowledgments

The editors acknowledge CRC Press for this opportunity and professional support. Our special thanks to Ms. Cindy Renee Carelli, Executive Editor, CRC Press—Taylor & Francis Group for the excellent support she provided in helping me to complete this book. Thanks also to Ms. Erin Harris, Editorial Assistant to Ms. Cindy, for her follow up and aid. We would also like to thank all the chapter authors and reviewers for their availability for this work.

Mangey Ram
Graphic Era (Deemed to be University), India

Om Prakash Nautiyal
Uttarakhand Science Education & Research Centre, India

Durgesh Pant
Uttarakhand Science Education & Research Centre, India

Editors

Mangey Ram received a Ph.D. degree majoring in mathematics and minoring in computer science from G. B. Pant University of Agriculture and Technology, Pantnagar, India. He has been a faculty member for around twelve years and has taught several core courses in pure and applied mathematics at undergraduate, postgraduate, and doctorate levels. He is currently a research professor at Graphic Era (Deemed to be University), Dehradun, India. Before joining the Graphic Era, he was a Deputy Manager (Probationary Officer) with Syndicate Bank for a short period. He is Editor-in-Chief of the *International Journal of Mathematical, Engineering and Management Sciences*, Book Series Editor with Elsevier, CRC Press—Taylor & Francis Group, De Gruyter Publisher Germany, River Publisher, USA, and the Guest Editor & Member of the editorial board of various journals. He has published more than 225 research publications in IEEE, Taylor & Francis, Springer, Elsevier, Emerald, World Scientific, and many other national and international journals and conferences. His fields of research are in the areas of reliability theory and applied mathematics. Dr. Ram is a Senior Member of the IEEE, life member of Operational Research Society of India, Society for Reliability Engineering, Quality and Operations Management in India, Indian Society of Industrial and Applied Mathematics. He has been a member of the organizing committee of several international and national conferences, seminars, and workshops. He has been conferred with the *Young Scientist Award* by the Uttarakhand State Council for Science and Technology, Dehradun, in 2009. He has been awarded the *Best Faculty Award* in 2011; *Research Excellence Award* in 2015; and, recently, the *Outstanding Researcher Award* in 2018 for his significant contribution to academics and research at Graphic Era Deemed to be University, Dehradun, India.

Om Prakash Nautiyal received the Ph.D. degree in physics from H.N.B. Garhwal Central University Srinagar, Garhwal, Uttarakhand, India. He also acquires Master's degree in three different subjects: physics, information technology and education. He has been a faculty member for around ten years in various capacities and has more than nine years' experience as a "Scientist" in Uttarakhand Science Education & Research Centre (USERC), Department of Science & Technology, Government of Uttarakhand. Dr. Om Prakash is on the editorial board of international and national journals of repute. He is a regular reviewer for several national and international journals and has also been a member of the organizing committee of a number of international and national conferences, seminars and workshops. He is working as Principal Investigator and Coordinator of some external funded projects funded by Govt. of India, BRNS, DAE; DST and MoES etc. He has authored/edited more than 5 books and published more than 30 publications in national & international reputed journals/book chapters/conference proceedings.

Durgesh Pant is among the earliest adopters and pioneers of computer science and information technology in the northern Himalayan region of India. He is credited with establishing the Department of Computer Science of Kumaun University, Nainital as early as 1989–1990. For about thirty years, he has been an avid promoter of computer science, IT and science & technology at all levels. He has chaired, participated in, and presented at conferences and seminars across the world. Given his penchant for leveraging science and technology, he has always been involved in developing scientific and technological solutions, innovations which are beneficial to the society at large and particularly to the "last-mile" learners in difficult geographies like Uttarakhand in India. A strong votary of Open and Distance Learning (ODL), he has been closely involved with it for over two decades, of which over a decade was spent with Indira Gandhi National Open University as coordinator. Since 2010, he has been working with Uttarakhand Open University as professor & Director of the School of Computer Science and Information Technology. In addition, he has also worked as the Director, Uttarakhand Space Application centre (USAC), Govt of Uttarakhand for about three years. Presently, he is also working with Uttarakhand Science Education and Research Centre (USERC), Department of Science & Technology, Govt. of Uttarakhand as its Director. Prof. Pant is also credited with establishing a unique centre for people with special needs at USERC. This centre is coming up as a hub for organizations working in the spheres of disabilities, particularly in the domains of leveraging science and technologies to this effect. Serving for nearly three decades in different fields has enriched Prof. Pant to have a balanced viewpoint. He has this conviction that Science, technology, environment, engineering and development all have to coexist in harmony. An MCA, PhD in Technology and an MBA, Prof. Pant has addressed seminars and conferences across the world and a number of research papers and books to his credit, in addition to innovations and patents.

Contributors

Prof. (Dr.) Anu A. Gokhale
Department of Technology
Illinois State University
Normal, Illinois, USA

Dr. Saurabh Kapoor
Department of Education in Science and Mathematics
Regional Institute of Education
Bhubaneswar, India

National Council of Educational Research and Training
Ministry of HRD
New Delhi, India

Dr. Sachin Kumar Mangla
Plymouth Business School
University of Plymouth
Plymouth, UK

Prof. (Dr.) Suresh Kumar Sharma
Centre for Systems Biology and Bioinformatics
Punjab University
Chandigarh, India

1 Develop Academic Writing Skills for a High Impact Research Project

Sachin Kumar Mangla

Contents

1.1 INTRODUCTION

"What is research?" is a question that every scientist/academician/researcher asks at some point in his career, but it is one that is very difficult to answer. With experience, however, each arrives at their own answer. The word "research" derived from an old French word, "*recerchier*," which means "to look for/to search for/search to search." Research may simply be defined as a "search over search." It is a search for the "What, Why, Why not, How, In what ways?" of the phenomenon, ideas, and facts of varied fields of sciences. It is a method for exploring, investigating and collecting information targeted at the discovery of new and novel facts for interpreting existing information, to discover or revise facts, theories, and applications (Kasi, 2009). The joy of research must be found in doing because every other harvest is uncertain (Nuttall, 1935). The new, novel and unknown aspect of research adds joy to the research, but it is also what makes it so difficult. However, this difficulty and complexity bring with it the encouragement and motivation to research. Thus, it is important to enjoy the process of research and to see the entire process as a learning experience.

The aim of the present chapter is to provide a brief guide for Early Career Researcher (ECRs) about how to begin and move forward effectively in their research fields, and how to publish their research in renowned journals. We start by explaining the basic research process, the structure and organization of a research paper, and the article with views of reviewers about the essentials of a good research paper for publishing in top journals.

To start with, the research process could be either inductive or deductive. Inductive research is a "bottom-up" research process, which is primarily a data-driven-exploratory process and is regarded as qualitative research. It uses methods such as case studies, grounded theory, and ethnographic study, paving the way for new theoretical explanations, and the development of new hypotheses in a field. The deductive process is a "top-down" approach, which focuses on verifying an a priori hypothesis. It places itself on the logical premises of a theory, where is a theory and its proposed hypothesis are correct then the results upon empirical or analytical testing must come as expected (Woo et al., 2017). In a nutshell, inductive research aims to analyze an observed phenomenon, whereas deductive research aims to verify the observed phenomenon. It is to be recognized that a balanced mix of deductive and inductive research is essential for the advancement of a field. While inductive qualitative research could be complex and laborious, especially for ECRs. But it is essential for new theoretical explanations in a field. A particular field can only achieve slight advances without the adoption of such an approach. Similarly, deductive research is essential to validate theories and verify principles across the field. Multiple factors play a crucial role in publishing both types of research.

An ECR should ideally decide his methodological approach based upon the theoretical positioning of his research. While a decision over the adoption of a particular research methodology is important, publishing the research in top-tier journals is still a different ball game. It is critical in terms of advancing an ECR's career as well as for promoting their field of research. It is often observed that a handful of researchers publish many papers in top-tier journals, whereas many researchers publish just a few papers over a long period. Straub (2009) proposes that one explanation for such a distribution of publications is that, for the majority of the time, people do not have a full understanding of why top journals accept a particular paper. Stout et al. (2006) identify poor motivation, poor design, and little original contribution to the field as the three most significant reasons for the rejection of an article. Thus, it can be said that "publishing is an art and a journey." A researcher needs to enjoy this journey and learn through this journey about how to publish his research in good journals.

1.2 STRUCTURING AND ORGANIZING A PAPER

Research is a time- and effort-consuming journey; thus, an ECR must be passionate and determined with regard to his research. An ECR has to begin from the start of this journey and must keep some key points in mind as he goes about this journey, as shown below.

- Start with a subject of interest;
- Narrow down to a topic;
- Close in on the tentative research questions and objectives;
- Form a preliminary bibliography;
- Start taking notes;
- Outline the paper;
- Write a rough draft;
- Edit your paper;
- Write a final draft.

While bearing these points in mind, it is important that the researcher should know how to structure and present his research effectively; thus, he must know the structure and organization of the paper, explained further here.

1.2.1 Abstract

The article starts with an abstract that briefly states the purpose of the research, motivation, research methodology, the principal results, and the major conclusions drawn from the paper. Depending upon the format of the journal, this can vary between 150 and 250 words. The abstract presented separately from the article is significant, since it is often the first section to be read and forms the first impression of the quality of the article. It should be understandable in its own right and should contain the most critical information about the article.

1.2.2 Introduction

The introduction is the first section of the main body of the paper. It should contain a brief outlining the central theme of the paper, the motivation driving the research, a short description of the work's objectives of the work and an adequate background of the research work done in the paper. Since the objective of this section is only to introduce and build a solid foundation for the sections to come, the introduction should avoid any detailed literature survey or a summary of the results. The research questions and objectives should be crisp and clearly mentioned. It is helpful to cite a few recent references from the targeted journal to showcase that the problem we are discussing is well within the scope of the publication. If any case/target industry is considered in the study, a brief description should be included. After introducing the research problem considered, the introduction section should present how the problem at hand has been solved, the methodology used, and the reasons for choosing the adapted methods/methodology.

In a broad sense, the introduction section should clearly state the nature of the problem? How is it formulated? What is the significance of the problem? How is it solved? After this, the last paragraph of the introduction should be regarding the organisation of the paper.

1.2.3 Literature Review

The literature review section is important to derive the theoretical placement of your research work. It identifies what other researchers have done so far in the field. It forms the basis of your study (Kasi, 2009). It is important that the researcher knows that they are not passing off something that has already been done in the field as his own novel contribution. Further, the literature review shows that the researcher is familiar with the existing available body of knowledge of his field, and has the capability to engage critically and make a contribution to it (ECREA European Media and Communication Summer School, 2011).

The literature review section should identify whether or not the previous literature has investigated a similar problem. It should identify what different perspectives have been presented on the problem, and what different methodologies have been used. Further, it helps us to identify the gaps in the literature and expand upon how we are making a contribution to it. Ideally, it should contain at least a few references from the target journal to identify how your work contributes to the journal.

1.2.4 Solution Methodology

This section should present a brief about the adopted research method/methods. Here, the focus shifts from broadly covering the methods used in field to the underlying theory that makes a particular method suitable for your research. It should provide the reasons for choosing a particular method and establish its suitability, over other methodologies, for your work. To present your logic for selection more firmly, this section can present a comparison of your adopted method with similar other methods, enlisting their advantages and disadvantages in addressing your research problem.

Upon establishing the suitability of the adopted method, the solution methodology should present a stepwise procedure of the methodology used to achieve the desired results. One should try to give a (brief) explanation these steps. It is helpful to provide a flowchart of the research framework, which can act as a blueprint of our solution methodology.

1.2.5 Data Analysis and Results

This section addresses a number of issues: How the data were collected? How the experts were chosen? The basis of their selection? Their background, experience, and other demographic details. The data analysis should provide the details of the steps taken to avoid a non-response bias, as well as the parameters that were considered in non-response bias. Further details about the measurement parameters should be explained. With regard to data collection, what scale was used to measure them? How were the data collected, i.e the mode of data collection? The collected data should be checked for its reliability, with the specification of the parameter to measure it. Finally, this section should provide a brief about the methods and statistical tests used for the data analysis and their results. The results here should only be explained in brief, with further discussed given in the next section.

1.2.6 Discussion of Findings

The findings of the research should be discussed here, as well as being validated with previous results. It should be identified how the results of the data analysis justify the proposed findings. Further, the findings should be in line with the research objectives while providing answers to the research questions proposed earlier. It should be clearly stated how the results and findings relate to the research objectives set forth in the article. Further, the findings of the research should be validated with previous research, any contradiction with the literature should be justified. To end this section, managerial implications of the research and the "take-home" information for the reader should be stated clearly.

1.2.7 Conclusion

The conclusion brings closure to the main body of the article. It should be brief and concise, focusing on the purpose of the work, and how the research objectives have been met. Remember that the conclusion should never introduce any new notion or aspect of your research. There should be no support in the literature; rather the whole discussion in this section should be grounded in the results and discussion section. The conclusion should list the limitations of the research, and how these limitations can be overcome in future researches. Finally, a brief about the implications and the scope for future research should be included.

1.2.8 References

References should always be in a uniform and standard format, in accordance with the target journal. It is important to have at least—three or four references from the target journal, of relevance to the research problem, to help build confidence among the editors and the reviewers about the scope of research. Cross-check the references twice, as any mistake in the references section leaves a bad impression with the editors.

1.3 ESSENTIALS FOR TOP JOURNALS

Research and PhD are long journeys. During a PhD, a scholar has to write up to 50000 words. Thus, regularity and discipline are key. One must write and write as much as possible. For publishing in top journals, one should always follow the top journals in his field of research. It is important to stay informed about the latest research trends in your field. In order to achieve this, you must know the top journals that are best suitable for publishing judging by the scope of your research, and should fully understand the requirements of those journals. It is advisable to use journal finder tools for this purpose. The worthiness of a research and if it is publishable in a high impact journal or not, is first governed by the novelty of the research and your contribution to the theory and practice in the field. It is important to highlight the novelty of the research with the utmost significance and should detail the possible application of your study. Furthermore, the article should be proofread for any language issues, flow, type errors, equations, figures, tables, etc. It is always advisable to take a hard copy of the research paper and read through it before submission.

1.3.1 Addressing the Reviewers

Revision is another important process of publication. It is important to respond to revisions with high priority and respect. Since it is not a paid job, you have to give due respect to the reviewer. But one must also not shy away from taking a stand. One must believe that revisions are doable, and must not be afraid of even lengthy revisions. While responding to the reviewers' queries, one must prepare a detailed response sheet and should highlight all the corrections and changes in the revised manuscript. Again, upon finalizing the revised manuscript, always take a hard copy of it and read it through.

1.4 CONCLUDING REMARKS

The article draws upon the difficulties faced by young researchers early in their careers. It sheds light on the steps involved in framing a research problem and structuring a research paper based on it. The presented steps can be replicated by researchers, who, with a combination of rigor and hard work, can target the top journals in their fields. It is a small step to prevent researchers from committing small mistakes that hamper their growth. In the future, such scientific guidelines could help authors to plan and design their research. The improvements in their manuscripts could maximize their publications as well as their knowledge contributions to the society and academia.

REFERENCES

ECREA European Media and Communication Summer School, University of Ljubljana, 2011. *Critical Perspectives on the European Mediasphere: The Intellectual Work of the 2011 ECREA European Media and Communication Doctoral Summer School.*

Kasi, P., 2009. *Research: What, Why and How?: A Treatise from Researchers to Researchers.* AuthorHouse, Bloomington. https://doi.org/10.13140/2.1.3274.1442

Nuttall, G.H.F., 1935. Theobald Smith, 1859–1934. *Obituary Notices of Fellows of the Royal Society* 1, 514–521. https://doi.org/10.1017/CBO9781107415324.004

Stout, D.E., Rebele, J.E., Howard, T.P., 2006. Reasons Research Papers Are Rejected at Accounting Education Journals. *Issues in Accounting Education* 21, 81–98. https://doi.org/10.2308/iace.2006.21.2.81

Straub, D.W., 2009. Editor's Comments: Why Top Journals Accept Your Paper. *MIS Quarterly* 33, 252–255.

Woo, S.E., O'Boyle, E.H., Spector, P.E., 2017. Best Practices in Developing, Conducting, and Evaluating Inductive Research. *Human Resource Management Review* 27, 255–264. https://doi.org/10.1016/j.hrmr.2016.08.004

Statistical Methods in Scientific Research

2

Suresh Kumar Sharma

Contents

2.1 DESCRIPTIVE STATISTICS

Statistics is a field of science concerned with the collection, compilation, summarization, analysis of data and then drawing valid inferences about the body of data when only a part of the data is observed. Statistical tools are employed in many fields—education, business, biology, medicine, psychology, agriculture, economics, just to mention a few. When the data to be analyzed are derived from biological sciences and medicine, we use the term biostatistics to distinguish this particular application of statistical tools and concepts. Descriptive statistics is a quantitative/qualitative summary of a particular data set. The concepts and methods necessary for achieving the objective of summarizing are usually presented in terms of numerical facts, or data, and in the form of tables or graphs. On the other hand, inferential statistics deals with the estimation and testing of a particular phenomenon under study. The nature of data can be broadly classified as:

- Structured data, and
- Unstructured data

Structured data can be put into numeric form such as height, weight, or age. Sometimes, the data are not exactly in numeric form, but can be easily converted to numeric. For example, gender is not numeric as such; it is male and female. But it can be converted into numeric form by assigning certain codes like male as 1; and female as 2. Similarly, socio-economic class can be classified: lower class as 1, middle class as 2 and upper class as 3. This type of data, which can be easily convertible into numeric form, also comes under the classification of structured data. However, data such as text, video, messages, emails etc., which cannot be directly put into numeric form, comes under unstructured data. Most of the statistical techniques can easily analyze the structured data; however, to analyze unstructured data we need the application of artificial intelligence, python, machine learning, deep learning, data science, and data analytics, along with Bayesian statistics.

Some of the words and phrases necessary for understanding the terminology of statistics are listed below:

Population: A population is the universal group from which data are sampled.

Sample: A sample is a part of a population, provided it represents the whole population.

Variable: A variable is a feature characteristic of any member of a population differing in quality or quantity from one member to another.

Qualitative variable: A variable differing in quality is called a qualitative variable or attribute, for example, eye color, gender, religion etc.

Quantitative variable: A variable differing in quantity is called quantitative variable, for example, age, height, weight etc.

Discrete variable: A discrete variable is the one which can take only certain isolated values, for example, number of members in a family, numbers of errors in a book etc.

Continuous variable: A continuous variable is one which can take all values within certain specified range, for example, age between 20 to 70 years.

Measurement scales: This may be defined as the assignment of numbers to objects or events according to a fixed set of rules. It is important to understand the role of various scales used in statistics. Here is the brief description about these scales.

Nominal scale: It is categorical in nature and there is no preference of one category over the other. Nominal data has no ordering and it is not metric. For example, gender, religion etc.

Ordinal scale: It is also categorical in nature but in this scale order is important. For example, severity of disease (mild, moderate, severe), socio-economic class (lower, middle, upper) etc.

Interval scale: In interval scale mathematical operations may not be valid. Addition or subtraction may be valid, but not multiplication and division. It is metric in nature. Moreover, zero has a meaning, for example, temperature, beauty scores etc.

Ratio scale: In ratio scale all mathematical operations (+, –, *, /) are valid. It is also metric in nature, for example, height, weight etc.

Nominal and Ordinal scales are used for categorical data and Interval and Ratio for measurement data. Of course, with some expertise, conversion of one scale to another is also possible.

Primary and Secondary Data: When data are used for the purpose for which they were originally collected, they are known as primary data and when the same data are used by someone else for any other purpose subsequently, they are termed secondary data. For example, if a company buyer obtains quotations for the price, delivery date and performance of a new piece of equipment from a number of suppliers with a view to purchase, then the data as used by the buyer are primary data. If these data are used later by the budgetary control department to estimate price increases of machinery over the past year, then it becomes secondary data.

Graphical representation of data: A graph is a method of presenting statistical data in visual form. A well-designed graphical presentation can effectively communicate the data's message in a language that is readily understood by almost everyone. The graphical methods for describing data are intuitively appealing. They can be used to describe either a sample or a population; quantitative or qualitative data sets. There are many varieties of graphs e.g. bar, multiple bar, stacked bar, line, pie, histogram, box-plot, error-bar, area etc.

Measures of Central Tendency: In many instances, we need to summarize the data by means of a single number, which is called the descriptive measure. These measures can be computed for either a sample or a population. If they are computed for a population, then they are known as parameters; if computed for samples, they are called statistics. Commonly used measures of central tendency are mean, median and mode. These measures tell how well a typical value represents the large list of values in the data set. It is useful for summarizing data and defines the center or middle of data.

Arithmetic Mean or Average: The mean is obtained by adding all the values in a sample or population and dividing by the total number of observations. It is denoted by $\bar{x}$, where

$$\text{Mean} = \bar{x} = \frac{\text{Sum of all the observations}}{\text{Total number of observations}} = \frac{\sum_{i=1}^{n} x_i}{n}.$$

Mean is a natural and most widely used measure of location; however, it can be oversensitive to outliers. In case of missing values or extreme values, mean is not the appropriate measure of central tendency as it carries the effect of inflation.

Median: After arranging the data either in ascending or descending order of magnitude, median is the middle most value which divides the finite set of values into two equal parts. If the number of observations (n) is odd, the median lies in the center, and for an even number of observations, it is the average of the two middle observations. The median is not drastically affected by extreme values. Sometimes, the median is also called positional average. It can be computed even for missing data and can also be located graphically.

Mode: The most frequently occurring value among all observations is called the mode. For example, if the set of values are 1,5,4,5,5,5,7,3,5,2 then the mode is 5, since this is the value which occurs most frequently. The mode also makes sense for qualitative and categorical data but the mean and median do not. It does not exist if each observation occurs exactly once. If some observations are repeated an equal number of times, then the distribution may have more than one mode. Unimodal distribution has one mode, bimodal distribution has two modes, and so on. The data with more than two modes are said to have multimodal distribution.

Measures of Dispersion or Variability: Dispersion means scatteredness. We study dispersion to have an idea about variability present in a set data. If all the values are same, there will be no dispersion. The amount of dispersion may be small, when the observations are close to each other and but it could be large, if they are widely scattered. For example, consider two series with following observations:

Series I: 1,5,9,13,17 (sum = 45, n = 5, $\bar{x} = 9$); **Series II:** 7,8,9,10,11 (sum = 45, n = 5, $\bar{x} = 9$).

Both the series are having 5 observations and have the same mean 9. Given the information that the mean of 5 observations is 9, we cannot make an idea about whether it represents Series I data or Series II data. Therefore, mean alone cannot describe the data completely, something is missing, and the missing part is dispersion. The range of Series I varies from 1 to 17 (R = 16), and that of Series II from 7 to 11 (R = 4). Thus, Series I is more dispersed than Series II, or we can say that the second series is more consistent than the first series. Commonly used measures of dispersion are Range and variance (or standard deviation). The square root of variance is called standard deviation.

Range: The difference between the highest and lowest values is called range, R = H – L. The usefulness of range is limited as it depends only on two observations, rather than the entire data set. These values may carry an inflation effect. The main advantage in using the range is the simplicity of its computation. However, it is a very crude measure of dispersion.

Variance: This is the most preferred measure of dispersion. It is based on the notion that when the values of a set of observations lie close to their mean, the dispersion will be less, and if they are far away from the mean, dispersion will be more. In computing the variance, we subtract mean from each observation, square the resulting differences, and add up the squared differences. This sum is then divided by sample size minus 1, to obtain the sample variance, and by sample size (n), for population variance.

$$\text{Population Variance} = \sigma^2 = \frac{1}{n}\sum_{i=1}^{n}(x_i - \bar{x})^2$$

$$\text{Sample variance} = s^2 = \frac{1}{n-1}\sum_{i=1}^{n}(x_i - \bar{x})^2$$

$$\text{Population Standard Deviation} = \sigma = \sqrt{\frac{1}{n}\sum_{i=1}^{n}(x_i - \bar{x})^2}.$$

$$\text{Sample Standard Deviation} = s = \sqrt{\frac{1}{n-1}\sum_{i=1}^{n}(x_i - \bar{x})^2}.$$

The variance represents squared units and it is not an appropriate measure of dispersion when we wish to express this concept in terms of the original units; therefore, the standard deviation is preferred as a measure of dispersion in original units.

Coefficient of Variation: When we want to compare the two or more given series in terms of standard deviation or variance, it may lead to fallacious results as they are measured in different units. In such situations, we use coefficient of variation (CV), which expresses the standard deviation as a percentage of mean. The formula is given by:

$$CV = \frac{\sigma}{\bar{x}} \times 100.$$

Since mean and standard deviation are measured in the same units, and they cancel out in computing the coefficient of variation. For sample observations, we replace σ by s (the sample standard deviation).

Partitioning Values (Percentiles and Quartiles): The mean and the median are special cases of a family of parameters known as location parameters. These descriptive measures are called location parameters because they can be used to designate certain positions on the horizontal axis when the distribution of a variable is graphed. For example, a distribution with a median of 150 is located to the right of a distribution with a median of 100 when the two distributions are graphed. Other location parameters include percentiles and quartiles. We may define a percentile as follows:

Definition (p^{th} percentile): Given a set of n observations, the pth percentile P is the value of a variable such that the proportion "p" of observations is less than or equal to P and the proportion (1 – p) of the observations are greater than P. Subscripts on P serve to distinguish one percentile from another. The 10th percentile, for example, is designated by P_{10} and the 70th by P_{70}. The 50th percentile, P_{50}, is the median. The 25th percentile is often referred to as the first quartile and it is denoted Q_1. The 50th percentile (the median) is referred to as the second quartile Q_2. Similarly, 75th percentile is referred to as the third quartile, denoted by Q_3.

Interquartile Range: A disadvantage of the range is the fact that it is computed from only two values, the largest and the smallest. A similar measure that reflects the variability among the middle 50% of the observations in a data set is the interquartile range.

The interquartile range (IQR) is the difference between the third and first quartiles: that is, IQR = $Q_3 - Q_1$.

Box and Whisker Plot: This is a very useful communicating device containing information about data. A Box is drawn between Q_1 and Q_3, the lower end of the box is placed at Q_1 and upper end at Q_3. Box is then divided into two parts by a horizontal line that aligns with median. It is generally drawn for comparison purpose. It is also useful to check the normality of the data.

The data are available for 169 patients between the ages of 18 and 69 years. The left-hand box-plot is drawn for the entire data and the right-hand box-plot to compare male (n = 91) and female (n = 78). The importance of the box-plot lies in the fact that it can identify outliers. Outlier is an observation which is away from the main body of the data. Box-plot identifies two types of outliers, probable (marked as "o") and extreme outliers (marked as *). In Figure 2.1 there is one probable outlier for male.

Box-plot is also used to compare value of a particular variable in two or more categories. In the above example, one can easily see that there is hardly any difference between the median ages of male and female categories, as the medians' lines are almost parallel. If the median lies in the center of the box, then we say that data follows normal distribution.

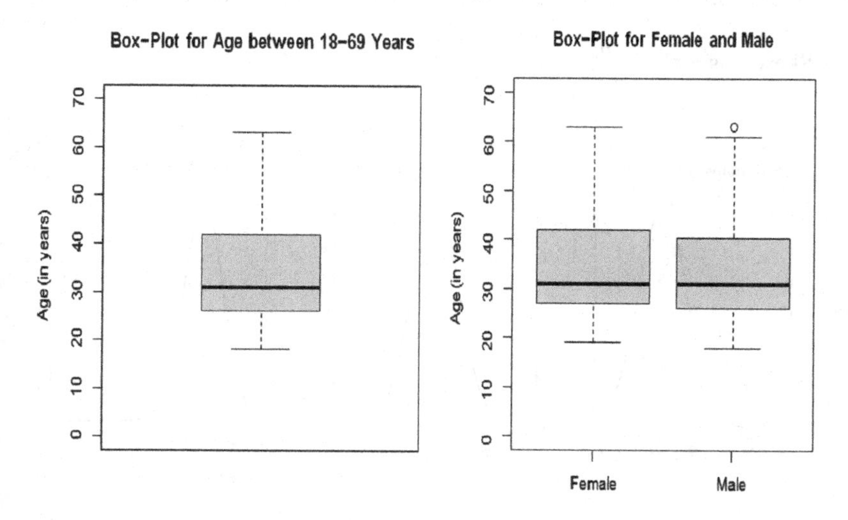

FIGURE 2.1 Box and Whisker plot for Age and Sex, and showing the outlier among Males.

Moments (Definition): The rth order central moment of a variable 'x', is defined as:

$$m_r = \frac{1}{n}\sum_{i=1}^{n}(x_i - \bar{x})^r, \; r = 0,1,2,.....$$

Note that $m_0 = 1$, $m_1 = 0$ always, and m_2 corresponds to the second-order moments, which is equal to σ^2, the variance, defined earlier. The m_3 and m_4 are called third- and fourth-order moments. These moments are helpful to determine Skewness and Kurtosis.

Skewness: This means "lack of symmetry." We study skewness to have idea about the shape of the curve which we can be drawn with the help of given data. A distribution is skewed if:

(i) Mean ≠ Median ≠ Mode.
(ii) Quartiles Q_1 and Q_3 are not equidistant from median (see Box-Plot).
(iii) The curve drawn with the help of given data is not symmetrical but stretched more to one side than to the other.

Figure 2.2 shows distribution for negatively skewed, symmetrical and positively skewed data.

Skewness can be measured either using absolute measures or relative measures, but the best measure of skewness is based upon moment defined as follows:

$$\text{Skewness} = \beta_1 = \frac{m_3}{m_2^{3/2}}.$$

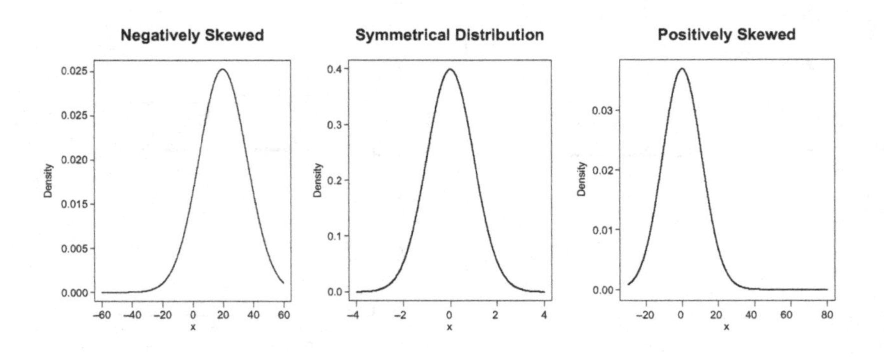

FIGURE 2.2 Distribution for negatively skewed, symmetrical and positively skewed data.

Where m_2 *and* m_3 are second- and third-order moments, computed from the given data. If

(i) $\beta_1 > 0$, the distribution is positively skewed.
(ii) $\beta_1 = 0$, the distribution is symmetrical.
(iii) $\beta_1 > 0$, the distribution is negatively skewed.

Kurtosis: It is a measure of peakedness of the curve. Even, if we know measures of central tendency, dispersion and skewness, still we cannot form a complete idea about the distribution, as it is clear from Figure 2.3.

Kurtosis is measured by coefficient β_2 or its deviation γ_2 given by

$$\text{Kurtosis} = \beta_2 = \frac{m_4}{m_2^2}, \quad \text{or} \quad \gamma_2 = \beta_2 - 3.$$

For the normal curve (also called the mesokurtic curve), $\beta_2 = 3$ or $\gamma_2 = 0$. The curve which is flatter than the normal is called platykurtic and for this curve $\beta_2 < 3$ or $\gamma_2 < 0$. Similarly, the curve which has a higher peak than the normal is called leptokurtic and for this curve $\beta_2 > 3$ or $\gamma_2 > 0$.

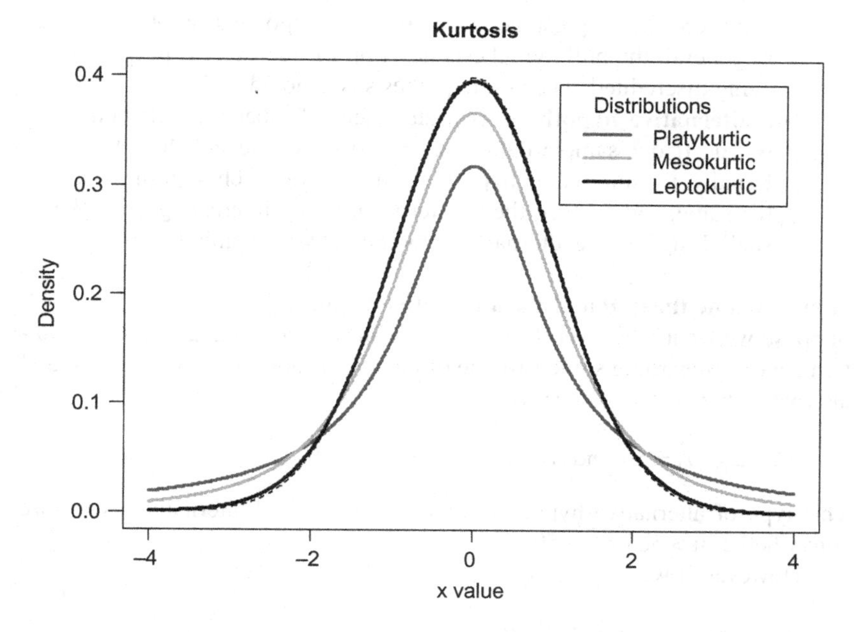

FIGURE 2.3 Distribution for Platykurtic, Mesokurtic and Leptokurtic data.

2.2 PARAMETRIC TESTS

Inferential statistical procedures generally fall into two possible categorizations: parametric and non-parametric. Depending on the level of the data you plan to examine, e.g. nominal, ordinal, continuous, a particular statistical approach should be followed. Parametric tests rely on the assumption that the data you are testing resembles a particular distribution (often a normal or "bell-shaped" distribution). Some basic concepts and definitions are required to understand the estimation and testing procedure. These concepts are discussed below:

Hypothesis: A tentative statement regarding the population relating to certain phenomenon in which the researcher is interested and wants to verify on the basis of a sample.

There are two statistical hypotheses (null and alternative hypothesis) involved in hypothesis testing, and these should be stated explicitly. The null hypothesis is the hypothesis to be tested. It is designated by the symbol H_0.

The **null hypothesis** is sometimes referred to as a hypothesis of no difference, since it is a statement of agreement with (or no difference from) conditions presumed to be true in the population of interest. In general, the null hypothesis is set up for the express purpose of being discredited. The null hypothesis is denoted by H_0.

The **alternative hypothesis** is a statement of what we will believe is true if our sample data cause us to reject the null hypothesis. Usually, the alternative hypothesis and the research hypothesis are the same, and, in fact, the two terms are used interchangeably. We shall designate the alternative hypothesis by the symbol H_1 or H_A.

How to frame the null and the alternative hypothesis?

Suppose we want to answer the question "Can we claim that a certain population's mean (regarding some variable of interest) is not 50?" Then the null and alternative hypothesis can be written as:

$$H_0 : \mu = \mu_0 = 50, \text{ and } H_1 : \mu \neq \mu_0 = 50$$

This type of alternative hypothesis is called two-sided alternative as we are sure whether $\mu > 50$ or $\mu < 50$.

However, if we are testing

$$H_0 : \mu = \mu_0 = 50, \text{ and } H_1 : \mu > \mu_0 = 50$$

This type of alternative hypothesis is called a one-sided alternative (right-tailed), and if

$$H_0 : \mu = \mu_0 = 50, \text{ and } H_1 : \mu < \mu_0 = 50,$$

then it is called a one-sided (left-tailed) alternative.

Depending upon the research question, one has to decide whether to use two-sided or one-sided alternative. If one is 100% sure that population mean would be greater (or less) than 50, then use one-sided, otherwise use two-sided.

Level of Significance: The significance level, also denoted by α, is a measure of the strength of the evidence that must be present in your sample before you will reject the null hypothesis and conclude that the effect is statistically significant. The researcher determines the significance level before conducting the experiment. The significance level is the probability of rejecting the null hypothesis when it is true. For example, a significance level of 0.05 indicates a 5% risk of concluding that a difference exists when there is no actual difference. Lower significance levels indicate that you require stronger evidence before you will reject the null hypothesis. It is generally fixed before conducting the experiment.

Type-I and Type-II Errors: No hypothesis test is 100% certain. Because the test is based on probabilities, there is always a chance of making an incorrect conclusion. When we do a hypothesis test, two types of errors are possible: Type I and Type II. The risks of these two errors are inversely related and determined by the level of significance and the power for the test. Therefore, we should determine which error has more severe consequences for the situation before you define their risks (Table 2.1).

TABLE 2.1 Type-1 and Type-II Errors

	TRUTH ABOUT THE POPULATION	
Decision based on sample	H_0 is true	H_0 is false
Accept H_0	Correct Decision (probability = 1 – α)	**Type II Error:** Accept H_0 when it is false (probability = β)
Reject H_0	**Type I Error:** Reject H_0 when it is true (probability = α)	Correct Decision (probability = 1 – β)

Test statistic: The statistics which is computed (based on the sample) and helps us take a decision either to reject H_0 or not to reject H_0.

Critical region or rejection region: The set of values for the test statistic that leads to rejection of H_0.

Critical values: The values of the test statistic that separate the rejection and non-rejection regions.

p-value: The p-value (or probability value) is the probability that the test statistic equals the observed value or a more extreme value under the assumption that the null hypothesis is true.

The logic of hypothesis testing is to reject the null hypothesis if the sample data are not consistent with the null hypothesis. Thus, one rejects the null hypothesis if the observed test statistic is more extreme in the direction of the alternative hypothesis than one can tolerate. The critical values are the boundary values obtained corresponding to the pre-set α level.

Confidence Intervals: The confidence level represents the percentage of intervals that would include the population parameter if we repeatedly take samples from the same population. A confidence level of 95% usually works well. This indicates that if you collected 100 samples, and made 100 95% confidence intervals, then one would expect approximately 95 of the intervals (out of 100) to contain the population parameter, such as the mean of the population.

Parametric Tests: Parametric tests make certain assumptions about a data set; namely, that the data are drawn from a population with a specific (normal) distribution. The majority of elementary statistical methods are parametric, and parametric tests generally have higher statistical power.

To make the generalization about the population from the sample, statistical tests are used. A statistical test is a formal technique that relies on the probability distribution, for reaching the conclusion concerning the reasonableness of the hypothesis. The **parametric test** is one which has information about the population parameter.

Assumptions of Parametric Tests: The observations must be independent (for example participants need to have completed the assignments separately, not in groups). The observations must be drawn from normally distributed populations. These populations must have the same variances (approximately).

How to check the normality of the data? There are three graphical procedures:

(i) **Histogram** (the shape of the histogram must be symmetrical).
(ii) **Box-and -Whisker Plot** (median line should be there in the center, Figure 2.1).
(iii) **Normal Quantile Plot** (Q-Q Plot) (quantiles of the observed data must be aligned to the quantiles of the standard normal distribution).

Commonly used Parametric Tests: (Notation: Mean = μ, Variance = σ^2, P = Proportion)

Z-Statistic (for large samples) is used for testing of

- Single population mean (e.g. H_0: $\mu = \mu_0$, where μ_0 is specified)
- Difference of two population means (e.g. H_0: $\mu_1 - \mu_2 = 0$ or H_0: $\mu_1 = \mu_2$)
- Single population proportion (e.g. H_0: $P = P_0$, where P_0 is specified)
- Difference of two population proportions (e.g. H_0: $P_1 - P_2 = 0$ or H_0: $P_1 = P_2$)

The basic requirement to apply Z-test is that the population variance must be known. Generally, the population variance is unknown and it restricts the application of Z-Statistic.

t-statistic (for small samples) is used for the testing of

- Single population mean (H_0: $\mu = \mu_0$, where μ_0 is specified)
- Difference of two population means (H_0: $\mu_1 - \mu_2 = 0$ or H_0: $\mu_1 = \mu_2$)
- Paired data (H_0: $\mu_d = 0$, where μ_d is the population mean difference before and after the treatment)
- Correlation coefficient (H_0: $\rho = 0$)

Note that t-statistic approaches to Z-Statistic for large samples. Hence, t-Statistic can be used for small as well as large samples.

Applications of Commonly used Parametric Tests with Examples:

Sampling from a Normally Distributed Population when the Population Variance is Unknown:

(i) **Single Sample Problem**

As we have already noted, the population variance is usually unknown in actual situations involving statistical inference about a population mean. When sampling is from an approximately normal population with an unknown variance, the test statistic for testing H_0: $\mu = \mu_0$, where μ_0 is specified, against H_1: $\mu \neq \mu_0$ is

$$t = \frac{\bar{x} - \mu_0}{s/\sqrt{n}}. \tag{2.1}$$

When H_0 is true, it is distributed as Student's t with (n – 1) degrees of freedom. The following example illustrates the hypothesis testing procedure when the population is assumed to be normally distributed and its variance is unknown.

Example 2.1 Subjects with medial collateral ligament (MCL) and anterior cruciate ligament (ACL) tears are considered. Between February 1995 and December 1997, 17 consecutive patients with combined acute ACL and grade III MCL injuries were treated by the same physician at the research center. One of the variables of interest was the length of time in days between the occurrence of the injury and the first magnetic resonance imaging (MRI). The data are shown in Table 2.2. We wish to know if we can conclude that the mean number of days between injury and initial MRI is not 15 days in a population presumed to be distributed normally.

We want to test the hypothesis

H_0: $\mu = 15$
H_0: $\mu \neq 15$

TABLE 2.2 Number of days until MRI for subjects with MCL and ACL tears

SUBJECT	*DAYS*	*SUBJECT*	*DAYS*
1	14	10	21
2	9	11	28
3	18	12	24
4	26	13	24
5	12	14	2
6	0	15	3
7	10	16	14
8	4	17	9
9	8		

From the given data mean ($\bar{x}$) = 13.2941, standard deviation (s) = 8.88654, n = 17, μ_0 =15 (under H_0).

Substituting these values in Equation (2.1), we get t = –0.791, p-value = 0.214.

Since p >.05, we do not reject the null hypothesis.

Conclusion: Based on these data, we conclude that the mean of the population from which the sample came is nearly 15.

(ii) **Two-Sample Problem**

Suppose we are interested to compare two populations with means μ_1 and μ_2 and wish to test the hypothesis H_0: $\mu_1 = \mu_2$, against H_1: $\mu_1 \neq \mu_2$. We take samples of sizes n_1 and n_2 from these two independent populations and compute their sample means $\bar{x}_1$, $\bar{x}_2$ and variances s_1^2, s_2^2, respectively. Then the following formula has *t-distribution* with ($n_1 + n_2 - 2$) degrees of freedom (d.f.) under H_0, for testing H_0 Vs H_1.

$$t = \frac{\bar{x}_1 - \bar{x}_2}{\sqrt{s^2\left(\frac{1}{n_1} + \frac{1}{n_2}\right)}}; \text{where } s^2 = \frac{(n_1 - 1)s_1^2 + (n_2 - 1)s_2^2}{n_1 + n_2 - 2} \quad (2.2)$$

Example 2.2: The Scientists examined subjects with hypertension and healthy control subjects. One of the variables of interest was the aortic stiffness index. Measures of this variable were calculated from the aortic diameter evaluated by M-mode echocardiography and blood pressure measured by a sphygmomanometer. Generally, physicians wish to reduce aortic stiffness. In the 15 patients with hypertension (group 1), the mean aortic stiffness index was 19.16, with a standard deviation of 5.29. In the 30 control subjects (group 2), the mean aortic stiffness index was 9.53, with a standard deviation of 2.69. We wish to determine if the two populations represented by these samples differ with respect to mean aortic stiffness index. The population variances are unknown, but are assumed to be equal.

It is given that $\bar{x}_1 = 19.16, \bar{x}_2 = 9.53$, $s_1 = 5.29$, $s_2 = 2.69$, $n_1 = 15$ and $n_2 = 30$.

Substituting these values in Equation (2.2), we get

t = 2.133, p=.023.

Since $p < 0.05$, we reject the null hypothesis.

Conclusion: On the basis of these results, we conclude that the two population means are significantly different.

(iii) **Paired-Sample Problem (paired t-test)**

In our previous discussion involving the difference between two population means, it was assumed that the samples were independent. A method frequently employed for assessing the effectiveness of a treatment or experimental procedure is one that makes use of related observations resulting from non-independent samples. A hypothesis test based on this type of data is known as a paired comparison test.

Related or paired observations may be obtained in a number of ways. The same subjects may be measured before and after receiving some treatment. Litter mates of the same sex may be assigned randomly to receive either a treatment or a placebo.

Instead of performing the analysis with individual observations, we use the difference (d_i) between pairs of observations, as the variable of interest. When the 'n' sample differences computed from the 'n' pairs of measurements constitute a simple random sample from a normally distributed population of differences, the test statistic for testing hypotheses about the population mean difference (i.e. H_0: $\mu_d = 0$, where μ_d is the population mean difference before and after the treatment) is:

$$t = \frac{\bar{d}}{S_d / \sqrt{n}}, \quad \text{where} \quad \bar{d} = \frac{\Sigma d_i}{n}, \text{ and } S_d = \sqrt{\frac{1}{n-1}\Sigma\left(d_i - \bar{d}\right)^2}.$$

Example 2.3: The data consist of the GBEF for 12 individuals, before and after fundoplication. We shall perform the statistical analysis on the differences in preop and postop GBEF. We may obtain the differences in one of two ways: by subtracting the preop from the postop percentages or by subtracting the postop from the preop percentages. Let us obtain the differences by subtracting the preop percentages from the postop percentages (Table 2.3).

TABLE 2.3 Gallbladder function in patients with presentations of gastroesophageal reflux disease before and after treatment

Preop (%)	22.0	63.3	96	9.2	3.1	50.0	33	69.0	64	18.8	0.0	34
Postop (%)	63.5	91.5	59	37.8	10.1	19.6	41	87.8	86	55.0	88	40

We want to test

$H_0 : \mu_d = 0$, against $H_1 : \mu_d < 0$.

From the given data, $\bar{d} = 18.075$, $S_d = 32.68$, n = 12. Hence,

$$t = 1.92, p = 0.041 \left(\text{one} - \text{tailed}\right)$$

Since $p < 0.05$, we reject H_0.

Conclusion: We may conclude that the fundoplication procedure increases GBEF functioning.

This statistic has t-distribution with (n – 1) degrees of freedom can be used for testing H_0: $\mu_d = 0$.

(iv) **Variance Ratio Test:** This is also called F-Statistic for testing equality of two population variances

$$\left(H_0 : \sigma_1^{\,2} = \sigma_2^{\,2}\right).$$

Decisions regarding the comparability of two population variances are usually based on the variance ratio test, which is a test of the null hypothesis that two population variances are equal. When we test the hypothesis that two population variances are equal, we are, in effect, testing the hypothesis that their ratio is equal to 1. For two populations we want to test the hypothesis

$$H_0 : \sigma_1^{\,2} = \sigma_2^{\,2}, \text{against } H_1 : \sigma_1^{\,2} \neq \sigma_2^{\,2} \text{ OR against}$$
$$H_1 : \sigma_1^{\,2} > \sigma_2^{\,2} \left(\text{OR } H_1 : \sigma_1^{\,2} < \sigma_2^{\,2}\right)$$

The statistics $F = \dfrac{S_1^2}{S_2^2}$ can be used for testing H_0 Vs H_1.

When the null hypothesis is true, the test statistic is distributed as F with numerator $(n_1 - 1)$ and $(n_2 - 1)$ denominator degrees of freedom.

Example 2.4: A Scientist compared meniscal repair techniques using cadaveric knee specimens. One of the variables of interest was the load at failure (in newtons) for knees fixed with the FasT-FIX technique (group 1) and the vertical suture method (group 2). Each technique was applied to six specimens. The standard deviation for the FasT-FIX method was 30.62, and the standard deviation for the vertical suture method was 11.37. Can we conclude that, in general, the variance of load at failure is higher for the FasT-FIX technique than the vertical suture method?

In this example, we want to test H_0: $\sigma_1^{\,2} = \sigma_2^{\,2}$, against H_1: $\sigma_1^{\,2} > \sigma_2^{\,2}$. It is given that

$S_1 = 30.62$, $S_2 = 11.37$, $n_1 = 6$ and $n_2 = 6$. Thus,

$$\text{Variance Ratio} = \text{V.R.} = F = \frac{S_1^2}{S_2^2} = \frac{(30.62)^2}{(11.37)^2} = 7.25$$

The p-value of this test is 0.019, which is less than 0.05. Hence, we reject H_0.

Conclusion: The failure load variability is higher when using the FasT-FIX method than the vertical suture method.

2.3 CORRELATION AND REGRESSION

In analyzing data for the health sciences disciplines, it is frequently desirable to observe and infer about the relationship between two numeric variables. For example, one may be interested in studying the relationship between blood pressure and age, height and weight, the concentration of an injected drug and heart rate, the consumption level of a nutrient and weight gain, the intensity of a stimulus and reaction time, or total family income and medical care expenditures, etc. The nature and strength of the relationships between variables such as these may be examined by correlation and regression analysis.

The word correlation is used in everyday life to denote some form of relationship or association. We might say that we have noticed a correlation between foggy days and attacks of wheezing. It is also assumed that the association is linear, that is, one variable increases or decreases a fixed amount for a corresponding unit increase or decrease in the other. The other technique that is often used in these circumstances is regression, which involves estimating the best straight line to summarize the association.

The degree of association is measured by a correlation coefficient, denoted by "r." It is sometimes called Karl Pearson's correlation coefficient after its originator and is a measure of linear relationship. The correlation coefficient is measured on a scale that varies from +1 through 0 to −1. Perfect correlation between two variables is expressed by either +1 or −1. When one variable increases as the other increases the correlation is positive; when one decreases as the other increases it is negative. Complete absence of linear correlation is represented by 0.

2.3.1 Scatter Diagram

When an investigator has collected two series of observations and wishes to verify whether there is a relationship between them, he or she should first construct a scatter diagram. The vertical scale represents one set of measurements and horizontal scale the other (Figure 2.4).

- If "r" is close to 1, we say that the variables are positively correlated. This means there is likely a strong linear relationship between the two variables, with a positive slope.
- If "r" is close to –1, we say that the variables are negatively correlated. This means there is likely a strong linear relationship between the two variables, with a negative slope.
- If "r" is close to 0, we say that the variables are not linearly correlated. This means that there is likely no linear relationship between the two variables, however, the variables may still be related in some other (like quadratic, cubic logarithmic etc.) way.

Mathematically, the strength and direction of a linear relationship between two variables is represented by the correlation coefficient. Suppose that there are n ordered pairs (x, y) that make up a sample from a population. The correlation coefficient r is given by:

$$r = \frac{n\sum xy - (\sum x)(\sum y)}{\sqrt{n\sum x^2 - (\sum x)^2}\sqrt{n\sum y^2 - (\sum y)^2}}$$

This will always be a number between –1 and 1 (inclusive).

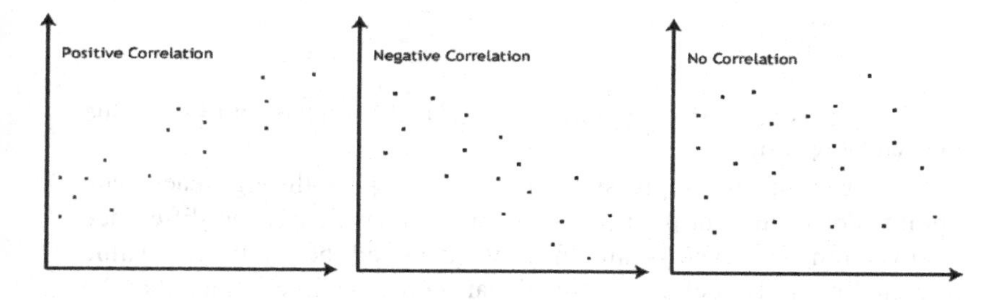

FIGURE 2.4 Scatter Diagram for positive, negative and no correlation between two variables

Example 2.5: The time X in years that an employee spent at a company and employee's hourly pay, Y, for five employees are listed in Table 2.4. Calculate and interpret the correlation coefficient r.

TABLE 2.4 Years of work (X) and hourly pay (Y) for employees at a company

X	Y	X_2	Y_2	$X*Y$
5	25	25	625	125
3	20	9	400	60
4	21	16	441	84
10	35	100	1225	350
15	38	225	1444	570
$\sum x = 37$	$\sum y = 139$	$\Sigma x^2 = 375$	$\Sigma y^2 = 4135$	$\Sigma xy = 1189$

Substituting these values with n = 5, in above formula, we get r = 0.97.

Interpretation: There is a strong positive correlation between the number of years an employee has worked and employee's salary, since "r" is very close to 1.

Linear Regression: If there is a significant linear correlation between two variables, the next step is to find the equation of a line that best fits the data. Such an equation can be used for prediction: given a new x-value, this equation can predict the y-value that is consistent with the information known about the data. This predicted y-value is denoted by $\hat{y}$. The line represented by such an equation is called the linear regression line.

The equation of a straight line is

$$\hat{y} = mx + b$$

where "m" is the slope of the line and "b" is the y-intercept (the y-value for which x is 0).

In general, the regression line will not pass through each data point. For each data point, there is an error, which is the difference between the observed y-value from the data and the predicted y-value by the line $\hat{y}$. By definition, this linear regression line is such that the sum of the squares of errors is the least possible. It turns out, given a set of data, there is only one such line where the slope "m" and y-intercept "b" are given by:

$$m = \frac{n\sum xy - (\sum x)(\sum y)}{n\sum x^2 - (\sum x)^2}, \; b = \bar{y} - m\bar{x},$$

where $\bar{x}$, and $\bar{y}$ are the mean values of x and y.

Example 2.6: The purpose of a study was to characterize acute hepatitis A in patients more than 40 years old. They performed a retrospective chart review of 20 subjects who were diagnosed with acute hepatitis A, but were not hospitalized. Of interest was the use of age (years) to predict bilirubin levels (mg/dl). The following data were collected (Table 2.5).

For these data m = 0.48, b = 9.42, r = 0.769 (reasonably good) and the equation of regression is given by:

$$\textbf{Bilirubin}(\textbf{mg/dl}) = \textbf{9.42} + \textbf{0.48} * \textbf{Age}(\textbf{in years})$$

Thus Bilirubin can be predicted for any Age from the above equation. 95% confidence intervals (or band) for regression line (also called line of best prediction) means that even if we take 100 such samples from the same population, 95% of the times the regression line will lie within this band and there is only a 5% chance that this may lie outside this band (Figure 2.5).

TABLE 2.5 Age and Bilirubin level for Hepatitis-A patients over 40 years' old

Age (Years)	78	72	81	59	64	48	46	44	42	45
Bilirubin (mg/dl)	27.5	22.9	24.3	18	16.1	12.9	11.3	9.8	7.8	10.8
Age (Years)	57	42	58	52	52	58	45	78	47	50
Bilirubin (mg/dl)	16.5	12.8	25.2	23.1	24.5	27.6	1.9	30.8	9.5	17.1

(Continued)

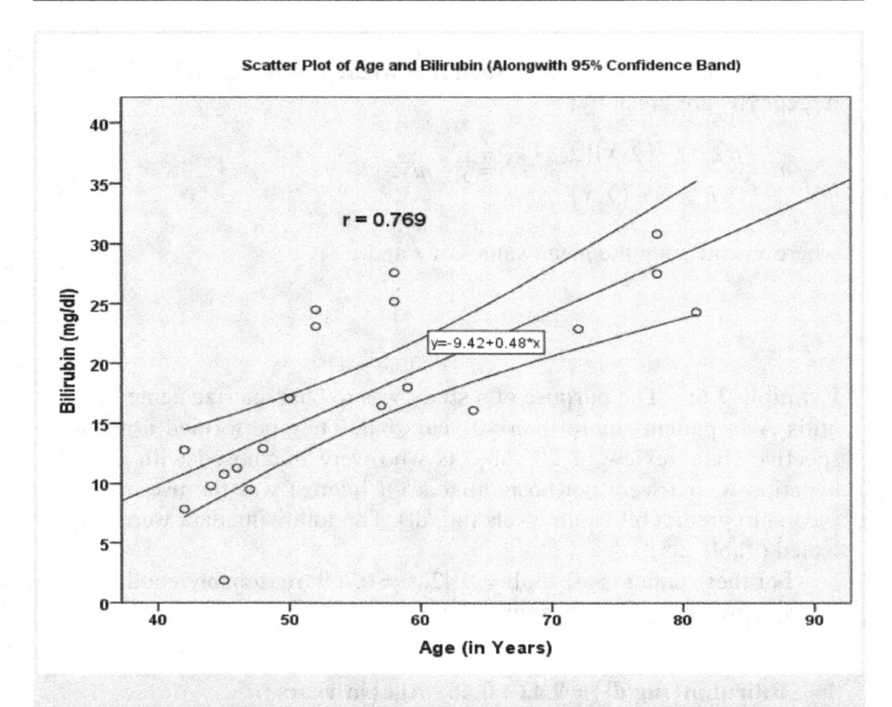

FIGURE 2.5 Scatter plot of Age vs. Bilirubin level for patients with Hepatitis A.

Above is an example of a situation where one can use Karl Pearson's correlation coefficient because both X and Y are measurements and the points are along a straight line. However, sometimes we come across a study where the data is in the form of ranks. In such situations, we use Spearman's correlation coefficient.

Spearman's Rank Correlation Coefficient:

Take a set of *n* points (x, y) and rank both x and y from 1 to n to get (r_x, r_y).

If the values of x and y are metric but non-normal, then it is advisable to compute Spearman's Rank correlation by replacing the original values of x and y in terms of ranks.

A correlation coefficient between r_x and r_y can be computed and it is easier to compute $d = r_x - r_y$, and then $r_s = 1 - \frac{6\sum d^2}{n\left(n^2-1\right)}$, which is the formula for Spearman's correlation coefficient.

Generally, this correlation is applicable if n is between 4 and 30. For really large n,

$z = r_s\sqrt{n-1}$ may be used.

Example 2.7: Five applicants for a job are rated by two officers, with the following results. Note that in this example the ranks are given initially. Usually the data need to be replaced by ranks (Table 2.6).

Note that $\sum d = 0$ and $\sum d^2 = 8$, and n = 5. Therefore,

$$r_s = 1 - \frac{6\Sigma d^2}{n(n^2-1)} = 1 - \frac{6(8)}{5((5)^2-1)} = 1 - \frac{2}{5} = 0.60.$$

The correlation is moderate and we can conclude that the opinion of the two raters may not be in agreement with each other.

TABLE 2.6 Ranks given to two applicants for a job

APPLICANTS	*A*	*B*	*C*	*D*	*E*
Rater-1	4	1	3	2	5
Rater-2	3	2	5	1	4

APPLICANT	r_x	r_y	d	d^2
A	4	3	1	1
B	1	2	−1	1
C	3	5	−2	4
D	2	1	1	1
E	5	4	1	1

Multiple Regression and Multiple Correlation

Multiple regression is a generalization of simple regression where we use more than one variable to predict Y. Suppose there are k predictors, say $X_1, X_2, \ldots, X_k$. For example,

If R^2 is the coefficient of determination for a regression

$$\hat{Y} = b_0 + b_1X_1 + b_2X_2 + \ldots b_kX_k \text{ then,}$$

R^2 is called the multiple correlation coefficient. Note that

$$R^2 = \frac{\Sigma(\hat{Y} - Y)^2}{\Sigma Y^2 - n\bar{Y}^2} = 1 - \frac{s_e^2}{s_y^2} \cdot \frac{n-k-1}{n-1},$$

where s_y^2 is the sample variance of y; s_e^2, is residual variance (e = $Y - \hat{Y}$), and that for large n,

$$R^2 \approx 1 - \frac{s_e^2}{s_y^2}.$$

Example 2.8: Suppose a Psychologist wants to estimate Intelligence Quotient (Y) based on the following predictors (independent variables).

X_1: *Mental age*
X_2: *Anxiety score*
X_3: *Depression score*
X_4: *Stress score*
X_5: *Communication score, and*
X_6: *Convolution score*
Y: Intelligent Quotient (IQ)

The data are given in Table 2.7.
The multiple regression equation can be written as:

$$\hat{Y} = b_0 + b_1 * X_1 + b_2 * X_2 + \ldots + b_6 * X_6$$

Where $\hat{Y}$ is the predicted value of Y. The coefficients b_0, b_1, b_2, …. . , b_6 are estimated from the given data. The following questions can be answered after carrying out the multiple regression analysis:

- Are all predictors (independent variables) important in predicting IQ, or are only a few of them relevant?
- If only a few of them are important, then which one is the most important, which the second most important, and so on.
- What is the coefficient of determination (R^2: called multiple correlation); that is, the strength of the relationship between Y and (X_1, X_2, … , X_6) taken together?

TABLE 2.7 Data related to psychological parameters

IQ	*MENTAL AGE*	*ANXIETY*	*DEPRESSION*	*STRESS*	*COMMUNICATION*	*CONVOLUTION*
105	47	85.4	1.75	5.1	63	33
115	49	94.2	2.1	3.8	70	14
116	49	95.3	1.98	8.2	72	10
117	50	94.7	2.01	5.8	73	99
112	51	89.4	1.89	7	72	95
121	48	99.5	2.25	9.3	71	10
121	49	99.8	2.25	2.5	69	42
110	47	90.9	1.9	6.2	66	8
110	49	89.2	1.83	7.1	69	62
114	48	92.7	2.07	5.6	64	35
114	47	94.4	2.07	5.3	74	90
115	49	94.1	1.98	5.6	71	21
114	50	91.6	2.05	10.2	68	47
106	45	87.1	1.92	5.6	67	80
125	52	101.3	2.19	10	76	98
114	46	94.5	1.98	7.4	69	95
106	46	87	1.87	3.6	62	18
113	46	94.5	1.9	4.3	70	12
110	48	90.5	1.88	9	71	99
122	56	95.7	2.09	7	75	99

The first step is to include all the independent variables, whether they are significant or insignificant (this option is available with Enter Method in SPSS). After applying this method, the following output is produced (Table 2.8)

TABLE 2.8 Model summary

MODEL	*R*	*R SQUARE*	*ADJUSTED R SQUARE*	*STD. ERROR OF THE ESTIMATE*
	0.998[a]	0.996	0.994	0.40723

Predictors: $X_1, X_2, \ldots, X_6$; Dependent: Y

Interpretation: The value of R-Square is very high ($R^2 = 0.996$), which means Y can be predicted from $X_1, X_2, \ldots, X_6$ with a coefficient of determination of 0.996. Since Adjusted R^2 is close to R^2, it implies that the data fit well to the multiple regression model. The significant ANOVA value indicates that at least one of the predictors is important to predict Y (there may be more than one) (Table 2.9).

TABLE 2.9 Analysis of variance (ANOVA)

Model	Sum of Squares	d.f.	Mean Square	F	p-value
Regression	557.844	6	92.974	560.641	0.0001
Residual	2.156	13	0.166		
Total	560.000	19			

After applying the Enter Method following output is produced (Table 2.10).

TABLE 2.10 Regression analysis output

Model	COEFFICIENTS				
	Unstandardized Coefficients		Standardized Coefficients	t-value	p-value
	B	Std. Error	Beta		
(Constant)	−12.870	2.557		−5.034	0.0001
Mental Age	0.703	0.050	0.324	14.177	**0.0001**
Anxiety	0.970	0.063	0.767	15.369	**0.0001**
Depression	3.776	1.580	0.095	2.390	**0.033**
Stress	0.068	0.048	0.027	1.412	0.182
Communication	−0.084	0.052	−0.059	−1.637	0.126
Convolution	0.006	0.003	0.038	1.633	0.126

Dependent Variable: IQ.

After careful examination, only three predictors are important (marked in bold), Mental Age, Anxiety and Depression (p-value is less than .001), while the other three variables, Stress, Communication and Convolution, are non-significant (p-value is more than .05).

Further, we need to apply step-wise regression so that only significant variables are retained in the final model. Moreover, step-wise regression will also identify which is most important predictor, the next one and so on.

The result of the step-wise regression produces the following output (Table 2.11).

Thus, only three predictors are retained in the final equation, with $R^2 = 0.995$, and Adj $R^2 = 0.994$. The predicted equation is

$$\hat{Y} = -13.667 + 0.906 * Anxiety + 0.702 * Mental\ Age + 4.627 * Depression$$

and the remaining insignificant variables have been removed from the model. This equation can be used to predict IQ for any given values of *Anxiety, Mental Age* and *Depression.*

TABLE 2.11 Step-wise regression analysis

MODEL SUMMARY (DEPENDENT VARIABLE=IQ)

Model	R	R Square	Adjusted R Square	Std. Error of the Estimate	Change Statistics				
					R Square Change	F Change	df1	df2	Sig. F (p-value)
1	0.950[a]	0.903	0.897	1.74050	0.903	166.859	1	18	0.000
2	0.996[b]	0.991	0.990	0.53269	0.089	175.162	1	17	0.000
3	0.997[c]	0.995	0.994	0.43705	0.003	9.255	1	16	0.008

COEFFICIENTS

Model	Unstandardized Coefficients		Standardized Coefficients	t-value	p-value	95.0% Confidence Interval for B	
	B	Std. Error	Beta			Lower Bound	Upper Bound
(Constant)	−13.667	2.647		−5.164	0.0001	−19.278	−8.057
Anxiety	0.906	0.049	0.717	18.490	0.0001	0.802	1.010
Mental Age	0.702	0.044	0.323	15.961	0.0001	0.608	0.795
Depression	4.627	1.521	0.116	3.042	0.008	1.403	7.852

[a] Predictors: (Constant), Anxiety
[b] Predictors: (Constant), Anxiety, Mental age
[c] Predictors: (Constant), Anxiety, Mental age, Depression.
Dependent Variable: IQ

2.4 CONCLUSION

In this chapter, some part of research methodology, including nature and types of data, has been covered. Descriptive statistics are used to describe the basic features of the data in a study. They provide simple summaries about the sample and the measures. Together with simple graphics analysis, they form the basis of virtually every quantitative analysis of data. Descriptive statistics help us to simplify large amounts of data in a sensible way. Each descriptive statistic reduces lots of data into a simpler summary. Measures of central tendency, dispersion, coefficient of variation, percentiles etc. are very helpful in the summarizing of data and all these have been discussed in this chapter. Inferential statistics use a random sample of data taken from a population to describe and make inferences about the population. Inferential statistics are valuable when the examination of each member of an entire population is not convenient or possible. The entire procedure is based on estimation and testing techniques, usually, involving parametric and non-parametric statistics. Some of the parametric tests have been discussed in this chapter. The goal of a correlation analysis is to see the strength of the relationship between two or more variables. Karl Pearson's and Spearman's Rank correlation coefficients with practical examples have been discussed. Regression analysis is a quantitative research method which is used when the study involves modelling and analysing several variables, where the relationship includes a dependent variable and one or more independent variables. Simple and multiple regression analysis techniques have been discussed in this chapter.

BIBLIOGRAPHY

1. Ambrosius W. T. (2007). *Topics in Biostatistics*, Springer.
2. Daniel W. W. (2010). *Biostatistics: A Foundation for Analysis in the Health Sciences*, Wiley Publication.
3. Rosner B. (2000). *Fundamentals of Biostatistics*, California: Duxbury Press.
4. Motulsky H. (1995). *Intuitive Biostatistics*, New York: Oxford University Press.
5. Brown M. B. and Forsythe A. B. (1974). Robust tests for the equality of variances, *Journal of the American Statistical Association*, 69, 364–367.
6. Kirkwood B. R. and Sterne J. A. C. (2003). *Essential Medical Statistics*, Oxford: Blackwell Science.
7. Bland M. (2001). *An Introduction to Medical Statistics*, Oxford University Press.
8. Bland M. and Altman D. G. (1986). Statistical methods for assessing agreement between two methods of clinical measurement. *The Lancet*, I, 307–310.

Teaching Learning Theories: *A Reflection*

Anu A. Gokhale

Contents

3.1 INTRODUCTION

Education has evolved differently in different countries and cultures according to the economic, political and social needs of society hugely influenced by the needs of the elite. In the United States, historically, the purpose of education has ranged from instructing youth in warfare and religious doctrine, to preparing them to live in a democracy with unequal educational opportunities during race-based segregation, to preparing workers for the industrialized twentieth-century workplace, to preparing professionals for twenty-first-century technology and service industry while continuously assimilating immigrants into mainstream society (Lucas, 1994; Banks, 2002a; Sloan, 2009).

In ancient India, education served two main goals—earning a livelihood and spiritual growth, although the latter trumped material knowledge (Choudhary, 2009). The concept, aims and ideals of education were correlated with the ideals of life. Individuals performed their duties with loyalty and responsibility; in general, people had an intellectual outlook in contrast to today's materialistic conception. The *gurukul*, a residential school run by a guru (or teacher) where all pupils received free education, was funded by the state and the public. Social interdependence was a deep and far-reaching support system. With invasions of India and its occupation by foreign powers, the country's educational system underwent significant changes, and today, it mostly mimics the western system.

This chapter begins with an overview of the current educational system largely influenced by twentieth-century teaching learning theorists and practitioners, discusses the educational system of ancient India, describes the goals of twenty-first-century education, and proposes a model for future-ready education.

3.2 REVIEW OF LITERATURE

Learning is generally defined as the acquisition of knowledge or skills through experience, study, or by being taught; it is a process that brings together personal and environmental experiences and influences for acquiring, enriching, or modifying one's knowledge, skills, behavior, attitudes, values, and worldviews. The review is limited to the teaching learning theories and pedagogies that have evolved in the West and those that existed in ancient India.

3.2.1 Teaching Learning Theories of the West

The field of education has been studied by innumerable philosophers, psychologists, and scientists who have focused on different aspects of the teaching learning paradigm. The intent here is not to provide an exhaustive or an all-encompassing review of literature but to identify major movements and theories in this space during the twentieth century, when the scientific study of learning was pursued in earnest. The major concepts discussed below include behaviorism, cognitivism, and constructivism.

3.2.1.1 Behaviorism

In behaviorism, it is assumed that a learner is essentially passive, responding to environmental stimuli. The learner behavior is shaped through positive reinforcement or negative reinforcement, which respectively increase or decrease the likelihood that the antecedent behavior will happen again. Learning is therefore defined as a change in behavior in the learner. It is based on the belief that all behavior is caused by external stimuli (operant conditioning), and can be explained without the need to consider internal mental states or consciousness. Among the key proponents of behaviorist theory are Pavlov (1849–1946), Thorndike (1874–1949), Watson (1878–1958) and Skinner (1904–1990); most of their work was done with animals and generalized to humans (Lefrancois, 1995; Ally, 2004).

According to this viewpoint, the teacher's role is to "give" information and is concerned with delivering knowledge with minimal regard for relating students' experiences to the subject matter. The purpose of assessment is to reward or punish, and consequently reinforce correct understanding.

3.2.1.2 Cognitivism

Cognitivism emerges after behaviorism and focuses on the thought process behind the behavior. Learning is viewed as an internal process and it is contended that the amount learned depends on the processing capacity of the learner, the amount of effort expended during the learning process, the depth of the processing, and the learner's existing knowledge structure. Although there are several proponents of constructivism, a few prominent names and the theories that they postulated are mentioned here.

Ausubel's (1963) *assimilation theory* distinguishes between *meaningful* and *rote learning*. Meaningful learning happens when new ideas get well anchored because they are related to and build upon pre-existing concepts. Rote learning, on the other hand, represents knowledge that remains unrelated and unanchored to existing concepts and is therefore easily forgotten. The *social cognitive learning theory*, advanced by Bandura (1978), proposes that people learn through observing others' behavior, attitudes, and outcomes of those behaviors. Cognitive load theory, introduced by Sweller (1988), differentiates between long-term memory and working memory, and proposes various ways of reducing extraneous cognitive load to enhance learning. Mayer (2002a) builds upon Sweller's work with cognitive theory for multimedia learning, which purports that the human working memory has two sub-components that work in parallel (visual and verbal/acoustic) and that learning can be more successful if both of these channels are used for information processing at the same time.

According to this viewpoint, the teacher must help the students relate the new concepts being taught to knowledge that the students already possess, so that new information may be better understood and remembered. Making multiple connections across concepts is critical to learning and the schema is reorganized in the learner's mind.

3.2.1.3 Constructivism

Constructivism builds upon cognitivism and claims that learners learn best when they can actively contextualize what they learn for immediate application and personal meaning. The learning is dependent on the situation itself, and the individual's purposes and active construction of meaning. Cognition is not just within the individual (as postulated by the cognitivists); rather, it is a part of the entire context—a distributed environment. John Dewey (1933–1998) is often cited as the philosophical founder of this approach, while Piaget (1970a) was an early advocate of the cognitive constructivist pedagogy; his theory argues that people produce knowledge and form meaning based upon their experiences. Bruner (1985) builds on the Socratic tradition of learning through dialogue, encouraging the learner to come to enlighten themselves through reflection. Vygotsky (1978) is a major theorist among the social constructivists, who cite the importance of social and motivational influences on cognitive development.

The constructivist viewpoint elevates the role of teachers from information transmitters to facilitators who assist the learner by contextualizing information. In the process, learners use a prior interpretation to construe a new or revised interpretation of the meaning of one's experience in order to guide future action. This takes the focus away from the teacher and puts it upon the student and their learning. Instead of asking and answering questions that only align with their curriculum, the facilitator must guide the student to come to the conclusions on their own instead of being told.

India's National Curriculum Framework 2005 (NCERT, 2005) requires that teachers have an in-depth knowledge of constructivist approaches. According to the National Curriculum Framework, "Constructing meaning is learning. ... Learners actively construct their own knowledge by connecting new ideas to existing ideas on the basis of materials/activities presented to them (experience)."

3.2.2 Pedagogical Approaches of the West

The pedagogy (or approach to teaching) translates into different instructional methods and classroom management strategies utilized by teachers. These are usually rooted in one or more of the theories of learning discussed in the previous section. This section discusses two of the most prominent instructional pedagogies: experiential learning and collaborative learning.

3.2.2.1 Experiential Learning

In direct contrast to the authoritarian, strict, didactic education, experiential learning is based on the belief that people learn best through experience. Accordingly, it emphasizes concrete experience to validate and test abstract concepts. In the USA, Dewey (1938) and Lewin (1890–1947) were early proponents of the value of hands-on learning (Harkavy & Hartley, 2010); this idea has later representations like the T-group movement and learning-style models (Kolb & Kolb, 2005; Boud et al., 1985; Miettinen, 2000). This approach is similar to authentic learning, which includes a wide variety of techniques focused on connecting what students are taught in school to real-world issues, problems, and applications. The overarching objective is to provide experience beyond the classroom in a real-world context.

3.2.2.2 Collaborative Learning

The term "collaborative learning" refers to an instruction method in which students at various performance levels work together in small groups toward the achievement of a common goal. The students are responsible for one another's learning as well as their own. Vygotsky (1978) contends that students are capable of performing at higher intellectual levels when asked to work together than when asked to work individually. Group diversity in terms of knowledge and experience contributes positively to the learning process. According to Slavin (1989), for effective collaborative learning, there must be "group goals" and "individual accountability." When the group's task is to ensure that every group member has learned something, it is in the interest of every group member to spend time explaining concepts to groupmates. Research has found consistently that students who gain most from cooperative work are those who give and receive elaborated explanations (Webb, 1987). Therefore, we ought to build in both "group goals" and "individual accountability" when using this pedagogy. According to Hagaman (1990), cooperative learning methods improve problem-solving strategies because the students are confronted with different interpretations of the given situation. The peer support system makes it possible for the learner to internalize both external knowledge and critical thinking skills and to convert them into tools for intellectual functioning.

3.2.3 Teaching Learning Theories of Ancient India

Education in ancient India was based on a sophisticated theory of knowledge derived from Indian philosophy and unique in that it included an understanding of the subject (the knower) in relation to the object (the known) and the process of knowing (Crozet, 2012). It placed a great emphasis on understanding the inner map of the subject/the person, the vehicle for learning, including the mechanics of thinking, the different layers of the mind and of a state beyond the mind, without denigrating the acquisition of objective or worldly/outer knowledge (Ramabrahmam, 2005).

Ancient Indian education was largely based on the Vedas, the Upanishads and several other texts. According to the Mundaka Upanishad, all knowledge is divided into two broad streams: (1) the *paravidya*, or the spiritual wisdom; and (2) the *aparavidya* or the secular sciences (Rama, 1990). A balanced combination of both is advocated, and was practiced in ancient times.

India has been a seat of learning since ancient times (Choudhary, 2009). It is often cited that the first university in the world was established sometime in the seventh century BC at Taxila (then in India and now in Pakistan). The original name of the place was "Takshashila," meaning "carved stone." A second university was founded in Nalanda in India in the fifth century BC and remained a center of excellence for learning for nearly 700 years, until it was destroyed in an invasion in the twelfth century. At its peak, Nalanda played host to nearly 10,000 students and about 2,000 world-renowned teachers, drawn from all over the globe. In spite of this success, very little is known about the learning theorists of those times. However, it is well-documented that at this time the education was completely free for everyone.

3.2.4 Pedagogical Approaches in Ancient India

In ancient India, learning happened through oral instruction as well as through experiential hands-on apprenticeships in a collaborative environment called a *Gurukul* (Mazumder & Bies, 1916). Gurukul is largely defined as a residential education system where pupils live together as equals near the guru, often in the same dwelling for a substantial number of years. A pupil leaves after the guru decides that the education is complete. At that time, the pupil offers *dakshina*, a token of gratitude, which may be in any shape or form, as desired by the guru.

According to Lall and Chowdhary (1952), the aim of education was the development of each individual. One of the basic philosophical assumptions was that the society is created for the benefit of an individual. Hence the content of education was decided by the interests of the pupil and whether it will result in individual development or not. As such, the content of education was not standardized to a great extent and there was sufficient flexibility in the curriculum to allow for free development.

3.2.5 Twenty-First-Century Education: Goals and Frameworks

According to Dede (2010), prominent conceptual frameworks for twenty-first-century education include the EnGauge by Metiri Group and NCREL (2003), three competency categories presented by the Organization for Economic Cooperation and Development (OECD), P21—Partnership for Twenty-First Century Skills (2006), and the Essential Learning Outcomes presented by the American Association of Colleges and Universities (AACU). The EnGauge framework (2003) includes digital age literacy, inventive thinking, effective communication, and high productivity skills. The OECD (2005) addresses competencies like using tools interactively, interacting in heterogeneous groups, and acting autonomously. The P21 (2006) emphasizes core subjects (e.g. English, math and science); twenty-first-century content (e.g. global awareness and entrepreneurial literacy); learning and thinking skills (e.g. creativity and innovation); information, communication and technology literacy; and life skills (e.g. adaptability and responsibility). The AACU (2007) add college-level essential learning outcomes to the P21 framework. A common strand among these frameworks is the instructional method—an inquiry-based collaborative approach that addresses real-world issues and questions.

3.3 THE FUTURE OF EDUCATION

An exploration of the future of education and learning has emerged from the concern about the inadequacy of the current system to meet the new demands of the twenty-first century, which is characterized as a knowledge- and technologically-driven era. In today's increasingly diverse societies, one of the principal goals of education is to learn how to live together in peace and harmony. It is critical that during individuals' formative years from childhood to youth, we develop their inner compass for humanitarian values, morals, and ethics and broaden narrow definitions of religions and beliefs. Going forward, education ought to address a learner's development of core subject knowledge, critical thinking, interpersonal and self-directional skills, and above all, humanitarian values.

The goal is to develop the learner's: (1) communication skills (through at least two languages, one of which must be the mother tongue); (2) logical and analytical thinking skills (through mathematics and science); (3) love and respect for physical work (through hands-on training in trades, physical education, sports, arts, and crafts); and (4) humanitarian values, morals, and ethics (through a study of philosophy and history and contemplation on the meaning of humanity). The teacher's objective is to teach a student how to self-learn while nurturing a spirit of comradery among learners.

The materials for self-study are made available to each learner through a technological device (even a printed book is a technological device). For teaching and learning methods, the learner is motivated through an emphasis on group discussions, raising and responding to questions, and hands-on learning experiences through team-based projects. A rationale for this methodology springs from western learning theories as well as Hindu texts. For example, the Bhagavad Gita is a dialogue and describes different paths to attain salvation; this indicates that "*one size does not fit all.*" The same format is true for some key Upanishads which are simply questions and answers. The message is that a learner must learn how to ask insightful questions in addition to learning to answer them.

After attaining basic literacy in a range of areas, the learner is now ready to pursue subject-specific study based on interest, aptitude, economic and social needs, and other personal factors. The system should be flexible enough to enable a learner to progress through competency levels and develop expertise in one or more areas. The teaching-learning methods would include self-reflection, peer-to-peer learning, and facilities for experimentation to test hypotheses. These techniques would further enhance the critical thinking skills of the learner.

3.4 SUMMARY

The purpose of education is to nurture individual aptitudes and strengthen their ability to: (a) make well-grounded decisions; (b) respect and work with people of different cultures and religious beliefs; and (c) foster individual growth and collective responsibility in a global community. Our current educational system is not built for collaborative learning; although students are learning together in the same classroom, the system has built-in competition among them through a ranking system, and their opportunities are limited by narrow performance measures. The solution is to rewire the closed educational system for an open and flexible model with twenty-first-century technologies.

REFERENCES

Ally, M. (2004). Foundations of educational theory for online learning. *Theory and Practice of Online Learning*, 2, 15–44.

American Association of Colleges and Universities. (2007). *College Learning for the New Global Century*. Washington, DC: AACU.

Ausubel, D. P. (1963). *The Psychology of Meaningful Verbal Learning*.

Bandura, A. (1978). Social learning theory of aggression. *Journal of Communication*, 28(3), 12–29.

Banks, J. A. (2002). Race, knowledge construction, and education in the USA: Lessons from history. *Race Ethnicity and Education*, 5(1), 7–27.

Boud, D., Keogh, R., & Walker, D. (Eds.). (1985). *Reflection: Turning experience into learning*. London: Kogan Page.

Bruner, J. (1985). Vygotsky: An historical and conceptual perspective. In J. Wertsch (Ed.), *Culture, Communication, and Cognition: Vygotskian Perspectives* (pp. 21–34). London: Cambridge University Press.

Choudhary, S. K. (2009). Higher education in India: A socio-historical journey from ancient period to 2006–07. *The Journal of Educational Enquiry*, 8(1) 50–72.

Crozet, C. (2012). The core tenets of education in ancient India, inspirations for modern times. *International Journal of Pedagogies and Learning*, 7(3), 262–265.

Dede, C. (2010). Comparing frameworks for 21st century skills. *21st Century Skills: Rethinking How Students Learn*, 20, 51–76.

Dewey, J. (1938). *Experiential Education*. New York: Collier.

Hagaman, S. (1990). The community of inquiry: An approach to collaborative learning. *Studies in Art Education*, 31(3), 149–157.

Harkavy, I., & Hartley, M. (2010). Pursuing Franklin's dream: Philosophical and historical roots of service-learning. *American Journal of Community Psychology*, 46(3–4), 418–427.

Kolb, A. Y., & Kolb, D. A. (2005). Learning styles and learning spaces: Enhancing experiential learning in higher education. *Academy of Management Learning & Education*, 4(2), 193–212.

Lall, I. D., & Chowdhary, K. P. (1952). *Principles and Practice of Education*. Delhi: Gulab Chand Kapur & Sons.

Lefrancois, G. R. (1995). *Theories of Human Learning: Kro's Report*. Pacific Grove, CA; Toronto: Brooks/Cole Publishing Company.

Lucas, C. J. (1994). *American Higher Education: A History* (p. 187). New York: St. Martin's Press.

Mayer, R. E. (2001). *Multimedia learning*. New York: Cambridge University Press.

Mazumder, N. N., & Bies, E. E. (1916). *History of Education in Ancient India*. Calcutta: Macmillan.

Metiri Group & NCREL. (2003). *EnGauge 21st Century Skills: Literacy in the Digital Age*. Chicago, IL: NCREL.

Miettinen, R. (2000). The concept of experiential learning and John Dewey's theory of reflective thought and action. *International Journal of Lifelong Education, 19*(1), 54–72.

NCERT. (2005). *National Curriculum Framework*. New Delhi: National Council of Educational Research and Training.

Organization for Economic Cooperation and Development. (2005). *The Definition and Selection of Key Competencies: Executive Summary*. Paris, France: OECD.

Partnership for 21st Century Skills. (2006). *A State Leader's Action Guide to 21st Century Skills: A New Vision for Education*. Tucson, AZ: Partnership for 21st Century Skills.

Piaget, J. (1970). *Science of Education and Psychology of the Child*. New York: Oxford University Press.

Rama, S., 1990, *Wisdom of the Ancient Sages: Mundaka Upanishad*. Himalayan International Institute of Yoga, Science, and Philosophy of U.S.A.; *99*, 139.

Ramabrahmam, V., 2005, Human cognitive process-An ancient Indian model, *Proceedings of the International Vedic Conference on Contribution of Vedas to the World*, Haridwar.

Slavin, R. E. (1989). *Effective Programs for Students at Risk*. Needham Heights, MA: Allyn and Bacon.

Sloan, W. M. (2009). Celebrating students' diverse strengths. *Supporting the Whole Child: Reflections on Best Practices in Learning, Teaching, and Leadership*, 255.

Sweller, J. (1988). Cognitive load during problem solving: Effects on learning. *Cognitive Science*, 12(2), 257–285.

Vygotsky, L. (1978). Interaction between learning and development. *Readings on the Development of Children*, 23(3), 34–41.

Webb, N. M. (1987). Peer interaction and learning with computers in small groups. *Computers in Human Behavior*, 3(3–4), 193–209.

4

Structured and Holistic Approach for Educational Research Proposal Development

Saurabh Kapoor

Contents

4.1 INTRODUCTION

Research design essentially refers to the plan or strategy of shaping the research: "design deals primarily with aim, purposes, intentions and plans within the practical constraints of location, time, money and availability of staff." Data Collection: Collection of data constitutes the first step in a statistical investigation. Utmost care must be exercised in collecting data because they form the foundation of statistical method. If the data are faulty, the conclusion drawn from them can never be reliable. The writing of a report is a presentation of facts and finding, usually as a basis for recommendations, written for a specific readership, and probably intended to be kept as a record. When some people write a report they do all of this in proper manner, but the really successful writers only spend a part of that time on writing until the end of the report. Before this they are planning their report—thinking about this purpose, and who is going to read it; deciding what to put into it, and shaping it accordingly. Even when they are finally writing it, they will probably spend just as much time thinking above how best to present their ideas as they do in actually putting them on paper. However, the educational research proposal development is viewed as more than the outcome of a formalized procedure. It might be a test of the researcher's ability to plan, conceptualize and achieve clarity. The researcher should ensure that their plan should be systematic. It is very important that a clear sense of direction must be there right from the beginning.

4.2 WHAT IS RESEARCH?

The goal of research is to improve the level of living in society. The word research carries an atmosphere of respect. As every object has both pros and cons, so does research. But the advantages of research have outnumbered the disadvantages of research and it has a place of its own in the field of study. In an academic environment, research activity is fivefold i.e., Master Dissertation; M. Phil. Dissertation; PhD Thesis; D. Litt. Thesis; and Assigned Research Project. The research projects are different from that of academic research degree with regard to their different timescales, resources and extent, pioneering qualities and rigor. A Research project actually involves groupwork on a pre-assigned topic by the funding agency; it has wide scope with regard to the greater resource availability.

4.3 DEFINITION OF RESEARCH

Research is composed of two words, "re" and "search," meaning to search again or a careful investigation to understand or re-examine the facts or to search for new facts or to modify older ones in any branch of knowledge. The term research is also used to describe an entire collection of information about a particular subject, but it is in general used by the students of higher schools. Research in common parlance refers to search for knowledge; one can also define research as a scientific and systematic search for pertinent information on a specific topic. Some people consider research to be a movement, a movement from the unknown to known. It is actually a voyage of discovery. Thus, research is an endeavor to discover, develop and verify knowledge.

P. M. Cook attributes the research taking the clue from each initial alphabets of the word **"research."**

R = Rational way of thinking;
E = Expert and exhaustive treatment;
S = Search for solution;
E = Exactness;
A = Analysis;
R = Relationship of facts;
C = Critical observation, Careful recording; Constructive attributes, and Condensed generalization.
H = Honesty and hard working.

The *Webster International Dictionary* defines research as "a careful critical enquiry or examination in seeking facts for principles, diligent investigation in order to ascertain something." The Advanced Learner's Dictionary of Current English, by contrast, lays down the meaning of research as "a careful investigation or inquiry specially through search for new facts in any branch of knowledge". J. W. Best opined that "research is not only specifically problem solving, but that it is also closely associated with verification of truth underlying the observed data." Thus, research is an intellectual act that begins with the asking of questions and progresses through the critical examination of evidence that is both relevant and reliable to the re-evaluation of the truth that is generalized and universal.

4.4 NEED OF RESEARCH

- To discover the truth, which is hidden and which has not been discovered as yet;
- To discover the solution of a problem;
- To expand the scope of theoretical knowledge;
- To discover the new application for old knowledge;
- To understand, analyze and explore the phenomena;
- To know the cause–effect relationship;
- To improve the level of living in society;
- For professional and intellectual development of the researcher by gaining knowledge;
- To obtain prestige and respect by a person or by the institution;
- To obtain a research degree; as a means of livelihood by way of obtaining the source of finance.

4.5 CHARACTERISTICS OF RESEARCH

- Research originates with a question or problem;
- Research requires a clear articulation of a goal;
- Research is guided by the specific research problem, question, or hypothesis or critical assumption;
- Research follows a specific plan of procedure;
- Research requires the collection and interpretation of data in attempting to resolve the problem that initiated the research;
- Research is, by its nature, cyclical; or more exactly, helical.

4.6 THE MEANING OF RESEARCH DESIGN

A research design is an arrangement of the essential condition and analysis of data in a form that aims to combine the relevance to research purpose with economy in the procedure. This means that a research design is not a highly specific plan to be followed without deviation. Rather, it is a series of signposts to keep one heading in the right direction. It is a decision regarding, what, when, how much, by what means concerning an inquiry or a research study contributes a research design. So a research design or a plan is tentative outline of the proposed research work. The plan is not a very specific one. It is simply a set of guidelines to keep the scholar on the right track.

- What is the study about?
- Why is the study being made?
- Where will the study be carried out?
- What type of data are required?
- Where can the required data be found?
- What will be the period of time required for study?
- What will be the sampling design?
- What are techniques used for data collection?
- How will the data be analyzed?
- Style of report writing.

4.7 NEED FOR RESEARCH DESIGN

Research design is needed because it facilitates the smooth sailing of the various operations, it makes the maximum information with minimum operations of expenditure, effort, time, and money. It is similar that before constructing a house we need the blueprint of it which is prepared by the experts (or architecture). Similarly, we need a research design or a plan in advance of data collection and analysis of our research project. Keeping in view the objectives of the research and the availability of staff, time and money, the preparation of the research design should be done with great care as any error in it may upset the entire project and research design.

Even then the need for a well-thought-out research design is at times not realized by many. The importance which this problem deserves is not given to it as a result many researchers do not serve the purpose of research which they had undertaken. In fact, they even may give misleading conclusions. There are some important points with regard to the need for research design:

- It may result in the desired type of study with a useful conclusion.
- It reduces inaccuracy.

- It helps to get optimum efficiency and reliability.
- It minimizes timewasting.
- It minimizes certain confusions, practical haphazard associated to any research problem.
- It helps in collection of data and research materials for tasking of hypothesis.
- It is a signpost for giving research a right direction.

4.8 FEATURES OF A GOOD RESEARCH DESIGN

A good design is often characterized by adjectives such as flexible, appropriate, efficient, economical and so on. The design which minimizes bias and maximizes the reliability of the data collected and analyses in consideration as good design. The design which gives the smallest experimental error is supposed to be the best design in many investigations. A research design appropriate for a particular research problem usually involves consideration of the following points:

- The objective of the problem to be studied.
- The nature of the problem to be studied.
- The availability of time money for the research work.

4.9 CHARACTERISTICS OF RESEARCH DESIGN

Regularity: State character of fact of being regular.
Verifiability: To ascertain text, the truth or accuracy of any opens for verification.
Universality: A state or quality of being universal or general.
Predictability: To predict or tell before with moderate accuracy.
Objectivity: Not subjective or unbiased.

4.10 RESEARCH PROPOSAL DEVELOPMENT OUTLINES

The Research Proposal is a complete description of the intended research, developed under the supervision of the assigned supervisor. Through the full proposal, the student needs to demonstrate convincingly that the study will make a contribution to a public health issue or problem. The full research proposal must be between 5 and 10 pages and should present the following:

- Title
  This should be brief, crisp, and communicate the intent of the study. The title reflects the brief outline of the work done inside it. The title must be holistic in nature.
- Brief Introduction
  This section may focus on need of the study in the Indian context, an overview of the work already done in the area and its linkage with the proposed study, and theoretical perspective (if any) to be followed. It will also include educational significance along with a rationale of the study. The introduction must have the planning to reach the research proposal. It must be clear in the introduction that the researcher has done some work in this direction.
- Background and statement of the problem (in the light of a thorough literature review)
  In this section it is to be certified that the work is being in the light of the literature review. This also have the problem formulation guidelines.
- Research question or hypothesis, aim and objectives
  The research questions are very important and an essential component of the research proposal. The development of the research can be certified from a research question or hypothesis. Specific achievable objectives or the corresponding research questions may be spelt out. The hypothesis, if any, may also be spelt out.
- Study design (type of study)
  This part shows the interest of the researcher about what kind of research is being undertaken. Is it qualitative or quantitative? Is it experimental or theoretical in nature?

- Study population and sampling
  A complete set of elements (persons or objects) that possesses some common characteristics defined by the sampling criteria established by the researcher. The population size is needed to be taken care by the researcher before study.
- Data collection methods and instruments
  This should include the details of the research design, the modality of collecting information and also the methodology of providing meaning to the collected information.
- Data analysis methods
  If applicable, statistical planning must be fully addressed, or the candidate should provide evidence that statistics are not required.
- Mechanisms to assure the quality of the study—e.g., control of bias, safe storage of data.
- Study period and time budgeting—timetable for completion of the project must be reflected.
- The proposal must include a statement about time duration in which the study is proposed to be completed. To make it more rational, it may be desirable to visualize various stages involved in the study vis-à-vis the time requirement for each stage. Resources required for the study, must also be included in budget if applicable
- Participants in the study or Organizational Framework—all people involved in the study, and the role they play, should be identified. An organizational chart indicating the tasks of the Principal Investigator (PI), Co-Principal Investigators (Co-Pis) (if any), and Junior Project Fellows (JPFs) (if any) with their duration should be given.
- Ethical considerations: conforming to accepted standards of conduct.
- References
- Appendices (copy of questionnaire, consent forms, etc.)

CONCLUSION

The objective of this chapter is to brief about the research proposal development in the systematic way. The research proposal development in the field of education is itself a challenge and demand. It is customary to conclude the research report with a very brief summary. The research must be more realistic and holistic in nature. The methodology must be effective and meaningful in nature. However, he expected findings make proposals more creative.

REFERENCES

Krathwohl, D.R. *How to Prepare a Dissertation Proposal: Suggestions for Students in Education and the Social and Behavioral Sciences*. Syracuse, NY: Syracuse University Press; 2005. pp. 45–47.

Labaree, R.V. *Organizing your social sciences research paper: Writing a research proposal*. AvailabLe from: http://www.libguides.usc.edu/writingguide

Marshal, J.G. Using evaluation research methods to improve quality. *Health Libraries Review* 12: 159–172, 1995.

Saleem, A., S.Z. Shabana Tabusum, and M. Sadik Batcha. Holistic approach of research work. *International Journal of Scientific and Research Publications* 4(7): 1–7, 2014.

Sudheesh, K., D.R. Duggappa, and S.S. Nethra. How to write a research proposal? *Indian Journal of Anaesthesia* 60(9): 631–634, 2016. doi:10.4103/0019-5049.190617

Index

Q

R

S

T

U

V

W

Z